I KNOW
NUMBERS

By Jon Welzen

Gareth Stevens
PUBLISHING

first concepts

I know numbers.
I see numbers!

I see 1 cupcake.

5

I see 2 hands.

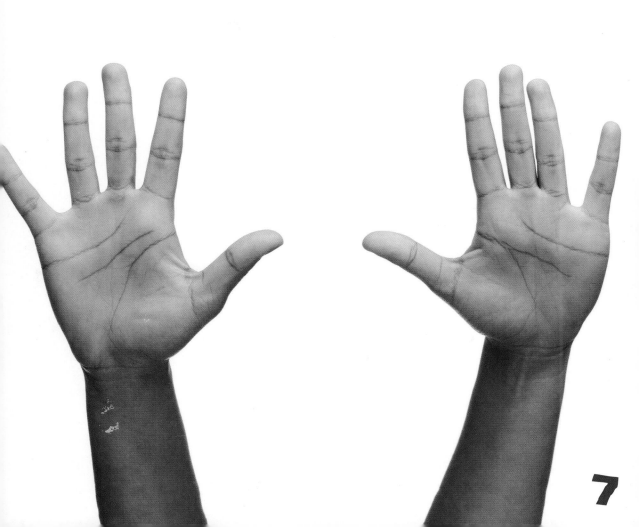

7

I see 3 friends.

I see
4 ice cream cones.

I see 5 blocks.

13

I see 6 rabbits.

15

I see 7 bears.

17

I see 8 frogs.

18

19

I see 9 balls.

20

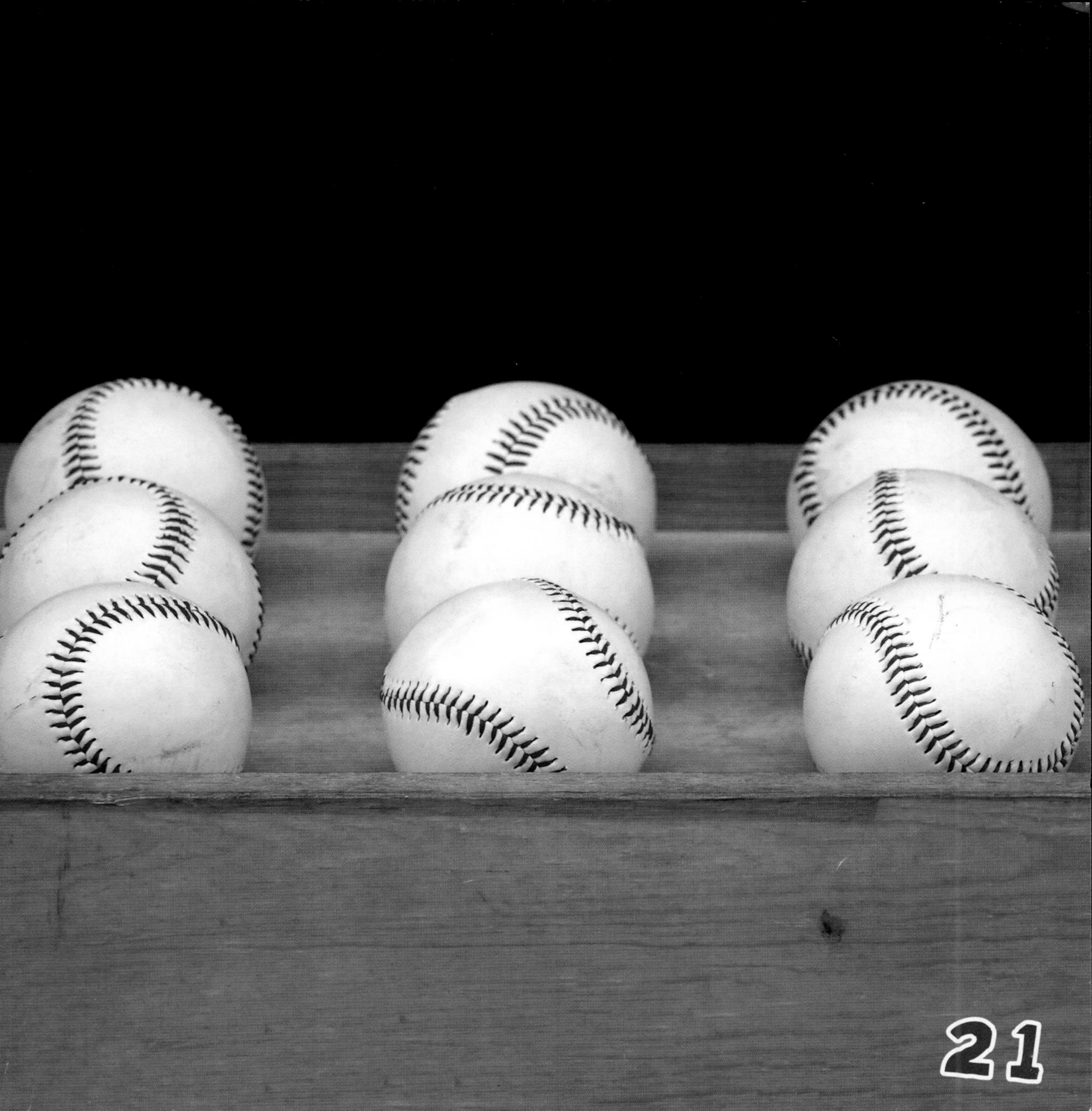

I see 10 toes!
What numbers
do you see?

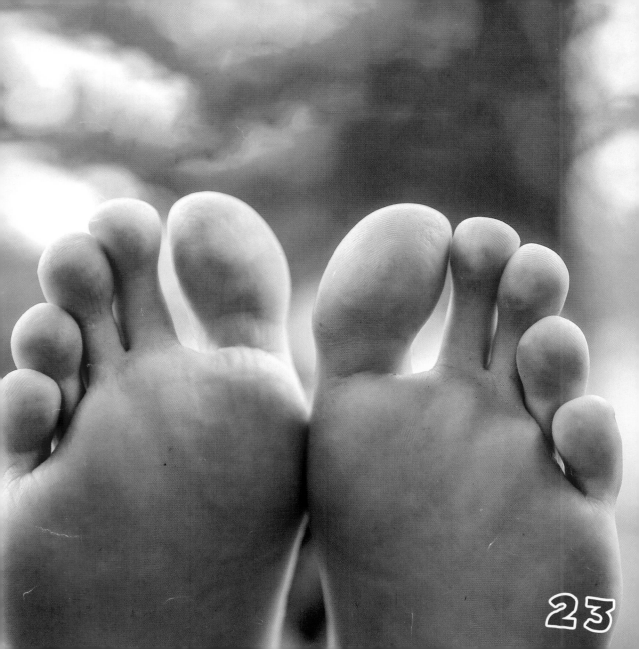

23

Please visit our website, www.garethstevens.com. For a free color catalog of all our high-quality books, call toll free 1-800-542-2595 or fax 1-877-542-2596.

Cataloging-in-Publication Data

Names: Welzen, Jon.
Title: I know numbers / Jon Welzen.
Description: New York : Gareth Stevens Publishing, 2017. | Series: What I know
Identifiers: ISBN 9781482454642 (pbk.) | ISBN 9781482454673 (library bound) | ISBN 9781482454659 (6 pack)
Subjects: LCSH: Numbers, Natural–Juvenile literature. | Counting–Juvenile literature.
Classification: LCC QA141.3 W45 2017 | DDC 513.2–dc23

First Edition

Published in 2017 by
Gareth Stevens Publishing
111 East 14th Street, Suite 349
New York, NY 10003

Copyright © 2017 Gareth Stevens Publishing

Designer: Sarah Liddell
Editor: Therese Shea

Photo credits: Cover, p. 1 (stripes) Eky Studio/Shutterstock.com; cover, p. 1 (numbers) sxpnz/Shutterstock.com; p. 3 Kdonmuang/Shutterstock.com; p. 5 Billion Photos/Shutterstock.com; p. 7 Syda Productions/Shutterstock.com; p. 9 Robert Kneschke/Shutterstock.com; p. 11 szefei/Shutterstock.com; p. 13 Maryna Kulchytska/Shutterstock.com; p. 15 Dmitry Kalinovsky/Shutterstock.com; p. 17 (main) Fesus Robert/Shutterstock.com; p. 17 (left front bear) harmpeti/Shutterstock.com; p. 19 jacotakepics/Shutterstock.com; p. 21 Shi Yali/Shutterstock.com; p. 23 Annette Shaff/Shutterstock.com.

Printed in the United States of America

CPSIA compliance information: Batch #CW17GS: For further information contact Gareth Stevens, New York, New York at 1-800-542-2595.